CHRONICLE BOOKS
SAN FRANCISCO

Mark Blum

acknowledgments

A work of this type is a collaborative effort based on the support, encouragement, and assistance of numerous persons. Among others, I am deeply indebted to: my wife and diving companion in the Galápagos, Marian; my friend and photo assistant on this project, Dan Blodget of Sub Aquatic Camera Repair Company; the able crew of the *Galápagos Aggressor II;* technical wizards Jack Mardesich, Ton Pennings, and Jacob Van Ekeren; wordsmith Mary Henness; the friendly staff of Fry Photographic; Jon Golden and Jan Burandt of 3D Concepts; Chuck Williamson of Williamson Engineering; Terry Schuller and Dave Reid of Ultralight Control Systems; Robert Kandel; Leo Cacitti; my editor, Steve Mockus; Robert Hessler, Scripps Institute of Oceanography; biologists Mary Wicksten and David Behrens; and Galápagos National Park naturalists Fernando Ortiz and Jaime Carrillo. I am especially grateful to sponsors Continental Airlines; the *Aggressor* Fleet; and Fuji Photo Film U.S.A., Inc.

Photo credits
Plate I: Robert Hessler, Scripps Institution of Oceanography
Plates 25 and 27: Dan Blodget

Library of Congress Cataloging-in-Publication Data:
Blum, Mark.
Galapagos in 3-D / Mark Blum.
p. cm.
ISBN 0-8118-3132-9
1. Zoology—Galapagos Islands. 2. Zoology—
Galapagos Islands—Pictorial works. I. Title.

QL345.G2 B58 2001
591.9866'5'0222—dc21 00-050932

Printed in Hong Kong

Designed by Azi Rad

Distributed in Canada by Raincoast Books
9050 Shaughnessy Street
Vancouver, British Columbia V6P 6E5

10 9 8 7 6 5 4 3 2 1

Chronicle Books LLC
85 Second Street
San Francisco, California 94105

www.chroniclebooks.com

ARCH
DARWIN · ISLAND
WOLF · ISLAND
PINTA
MARCHENA
GENOVESA
SOUTH AMERICA
GALÁPAGOS ISLANDS
GALÁPAGOS · ISLANDS
EQUATOR
PUNTA · VICENTE · ROCA
SANTIAGO
BANKS · BAY
COUSIN'S · ROCK
BARTOLOME
FERNANDINA
N. SEYMOUR
BALTRA
PINZON
SOUTH · PLAZA · ISLAND
DARWIN
RESEARCH · STATION
SANTA · CRUZ
SANTA · FÉ
ISLA · LOBOS
NORTH
SAN CRISTOBAL
ISABELA
FLOREANA
ESPAÑOLA

preface

The Galápagos Islands form an archipelago in the Pacific Ocean off the coast of Ecuador. Early human visitors called them *Las Islas Encantadas* ("The Enchanted Islands") because they thought the islands were free-floating, unconnected to the ocean floor. When Charles Darwin visited the islands in 1835, he was awed by their primal volcanic landscape and amazed by the number and variety of native creatures. In his journal, he wrote that the archipelago was like "a little world in itself," and that in this unique world, "we seem to be brought somewhat near to that great fact—that mystery of mysteries—the first appearance of new beings on this earth." His observations on the islands were central to his developing theories of evolution and natural selection, laid out 20 years later in his *Origin of Species*, which changed the way we understand the history of life on Earth.

From the very bottom of the sea floor, where the islands are born of fire from the Earth's core, to the mist-shrouded summits of their collapsed volcano peaks, the story of these unique islands is that of a place apart from the rest of our planet. Indeed, the astounding process of evolution is so clearly evident on these islands largely because they have never been connected to the mainland, or to each other. Animals have evolved to survive in unique habitats in relative isolation. Even under water, it is estimated that approximately 23 percent of Galápagos' sea life is composed of native species found nowhere else.

The Galápagos Islands have a combined landmass roughly the size of the state of Connecticut—about 5,000 square miles (8,000 square kilometers)—almost all of it incorporated in the Galápagos National Park, which includes a marine reserve. The park collaborates with the international, nonprofit Charles Darwin Research Station to manage the islands and their resources, focusing on environmental education, conservation, and captive breeding programs. Preservation efforts face considerable challenges, however, including tourism, fishing, the intrusion of non-native species, and the islands' ever-increasing human population. On the bright side, captive breeding programs and the removal of introduced species are giving native animals a fighting chance at survival, and it has been reported that the regular presence of scuba diving charter boats at the remote outposts of Wolf and Darwin Islands has helped to curtail illegal shark fishing in recent years.

The stereoscopic images included here are windows into this extraordinary world, offering the chance to explore its unique terrain and meet its unusual inhabitants up close in a way that standard, two-dimensional photographs simply can't match. None of the 3-D photographs in this book were created or enhanced with computers; all were taken with stereo cameras.

While this book is not meant to be comprehensive, I've tried to include the creatures and terrain that you'd likely see on a typical visit to the islands as well as images, such as the marine life on the islands' volcanic underwater slopes, that few visitors are lucky enough to encounter. I've included the scientific name (in parentheses) and the local Spanish name (in *italics*) for each creature. I sincerely hope that you will share the sense of wonder and fascination I have felt in these enchanted islands.

—Mark Blum

Plate I

Ocean Vent

Galápagos Rift

This unique stereo photograph was taken on the deep ocean floor along the Galápagos Rift, about 200 miles (120 kilometers) west of the Galápagos Islands. Here, at a depth of approximately 8,250 feet (2,500 meters), huge tectonic plates are spreading apart. Into the near-freezing and perpetually black water, deep-sea hydrothermal vents spew superheated minerals at temperatures of more than 700 degrees Fahrenheit (380 degrees centigrade). Although there is no sunlight for photosynthesis, animals have evolved to live by chemosynthesis, eating bacteria that feed on sulfur compounds expelled from the earth's crust. This photo, taken from a submarine, shows yellowish mussels and two types of crabs (brachyuran and galatheid) adapted for life in this harsh environment. The existence of this remarkable biological community without sunlight has led scientists to postulate that life may exist elsewhere in our solar system, such as on the moons of Jupiter.

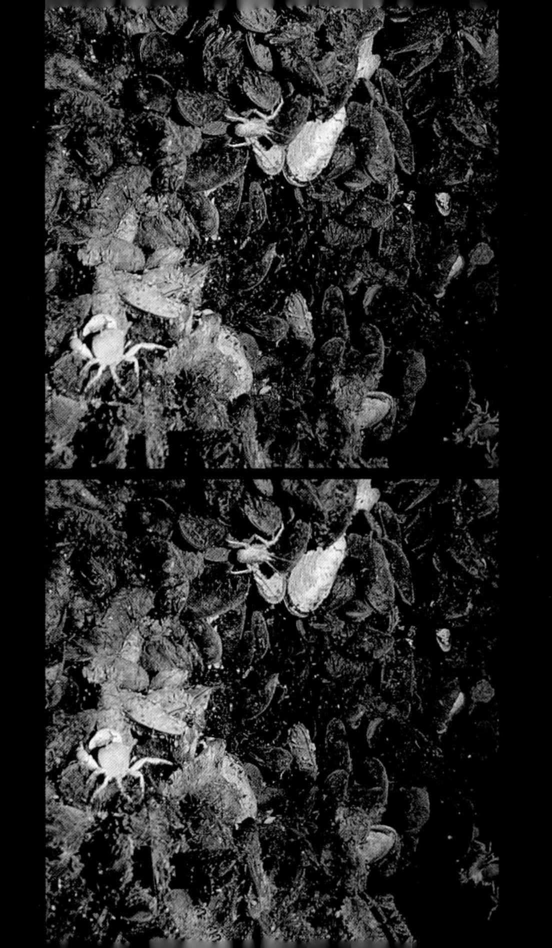

Plate 2

Darwin's Arch

Darwin Island

This monumental stone arch rises above a shallow underwater ridge extending from the east end of Darwin Island. Although named after Charles Darwin, the island was never visited by its namesake. The northernmost island in the archipelago, it is formed by the eroded summit of a large caldera, or collapsed shield volcano. Rising approximately 6,600 feet (2,000 meters) from the sea floor, the island and arch are composed of tuff (rock composed of fine volcanic detritus) approximately two and a half million years old. The arch is inhabited only by boobies and terns. Beneath the ocean's surface, however, teems a riot of marine life attracted to the swift current that splits around the arch. Experienced scuba divers look for a protected pocket in the submarine terrain, hold on, and simply let the underwater parade sweep by.

Plate 3

Scalloped Hammerhead Sharks

Sphyrna lewini · *Tiburón Martillo*

Wolf Island

Due to their uniquely flattened heads and dangerous reputation, hammerhead sharks are perhaps the most widely recognized of all shark species, but much about these animals remains mysterious. No one knows for sure why the head is so strangely configured, with eyes and nostrils on either lobe. Some of the hypotheses about this adaptation's benefits include better vision, sense of smell, and other sensory ability; and increased lift and maneuverability. Lateral lines of fluid-filled sensory canals, which run along the sides of the shark's head and body, aid in balance and hearing. Skin pores called ampullae of Lorenzini, located on the flattened head, contain receptor cells that can detect faint electrical fields emitted by their prey. The hammerhead sharks at Wolf and Darwin Islands are abundant. Patrolling individually and in schools at shallow depths, they seem not to be easily frightened by scuba divers.

Plate 4

Pacific Bottlenose Dolphins on Bow Wave

Tursiops truncatus *Delfín*

Wolf Island

Bottlenose dolphins frequently ride the bow waves of boats approaching or leaving Wolf and Darwin Islands. Known for their fast, powerful swimming abilities, bottlenose dolphins have been reported at bow-riding speeds faster than 35 miles per hour (55 kilometers per hour). The variety of bottlenose dolphin that keeps close to shore forms groups ranging from two to eighteen individuals, while herds that swim farther offshore can exceed 500 dolphins. Bottlenose dolphins sometimes mix with other cetacean groups, including porpoises, humpback dolphins, and some of the smaller species of toothed whales. Bottlenose dolphins will feed on almost anything they can catch and swallow. Many prefer to bite a fish in half, or remove its head, before swallowing it.

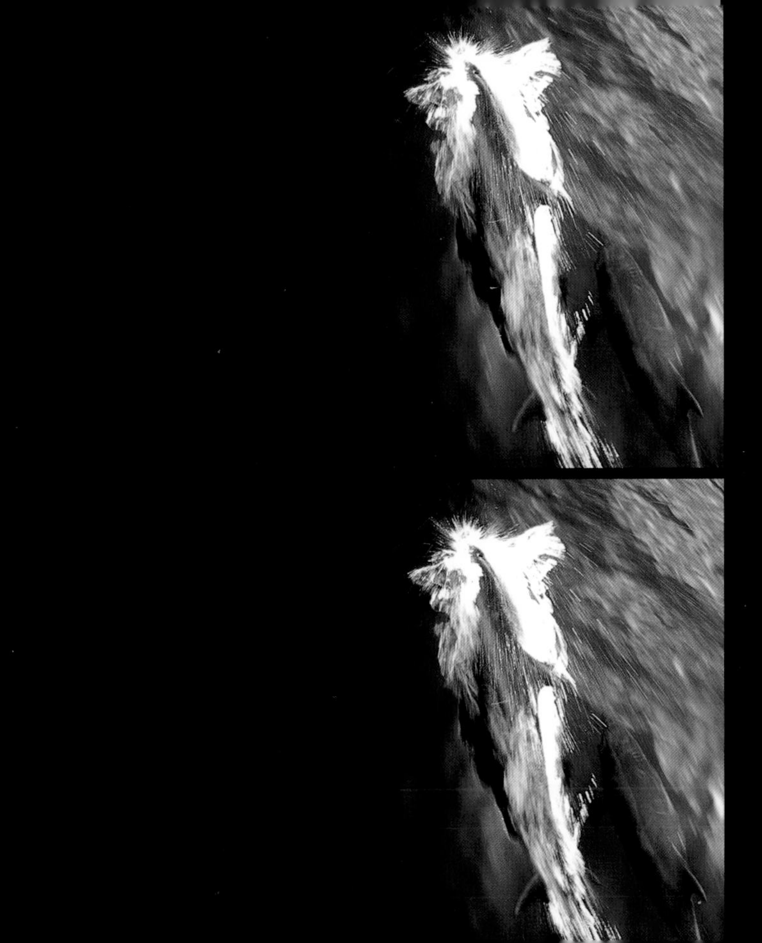

Plate 5

Whale Shark

Rhincodon typus *Tiburón Ballena*

Darwin Island

The astonishing whale shark may grow to over 55 feet (17 meters) long, making it the world's largest living fish. Surprisingly, little is known about these docile giants. It was only established in 1966 that female whale sharks give live birth (meaning they are ovoviviparous) rather than laying eggs. These gentle creatures are one of the few shark species that are filter feeders, eating plankton, small fish, squid, and shrimps. The whale shark feeds at or near the surface of the sea, where it inhales water and drains it through its massive gills, retaining the food in the process. They are not shy and, if unmolested, will often continue feeding even with divers nearby, allowing for close observation.

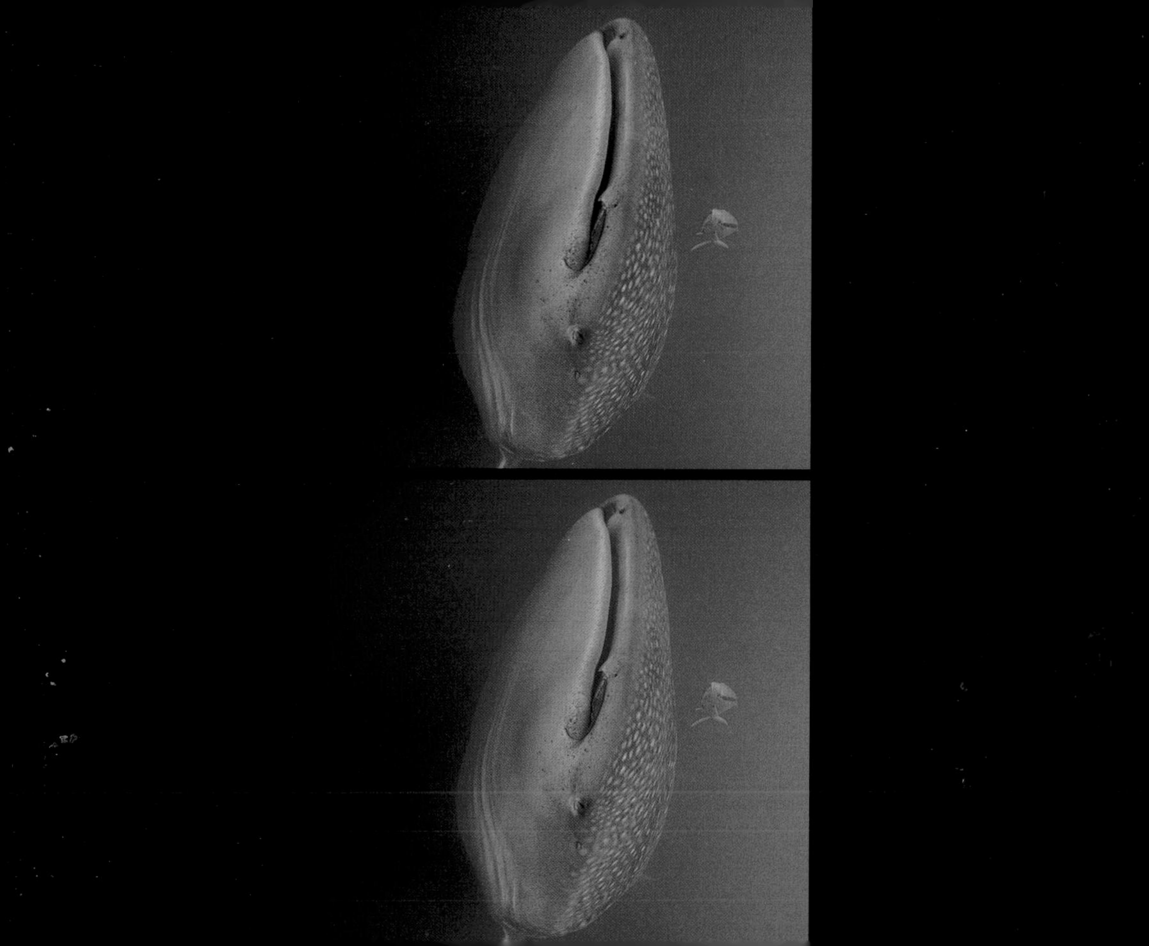

Plate 6

Swallowtail Gulls

Creagrus furcatus *Gaviotas Colabifurcada*

North Seymour Island

The native swallowtail gull is the world's only nocturnal gull. It feeds at night, 10 to 20 miles (6 to 12 kilometers) offshore, searching for squid and fish near the surface of the sea. This bird is endemic, meaning that it only lives in the Galápagos Islands. Its large, red-ringed eyes provide the gull with excellent night vision. Hunting at night lets the gull avoid competition with seabirds such as the red-footed booby that feed offshore during the day. The pair shown here is in courtship. Swallowtails construct nests of coral and lava pebbles on rocky cliffs and shores. Females lay a single, spotted egg, and both parents care for the offspring, which matures and leaves the nest along with the parents after about 90 days. They will all return in four to five months to start another breeding season, which lasts as long as conditions are right for rearing chicks.

Plate 7

Lava Gull

Larus fuliginosus *Gaviota de Lava*

North Seymour Island

The sooty-gray color of the lava gull is designed to camouflage it on the black lava shores of its native volcanic islands. This coloration may have developed not as a defense against predators, but to gain an advantage over competing scavengers. Only adult specimens, seen here, develop a white pattern around the eyes. Lava gulls eat sea lion afterbirth, the remains of fish left behind by sea lions, and worms, as this one is fishing for. They're also known to feed on newly hatched marine iguanas, Sally Lightfoot crabs, and eggs of various animals. Solitary by nature, lava gull pairs nest miles away from other gulls. Unlike the cliff-nesting swallowtail gull, lava gulls nest in vegetation near sheltered water. They are highly protective of their nests, dive-bombing with a sharp cry any intruder that comes near. Some estimates count only 200 pairs of these unique birds in existence.

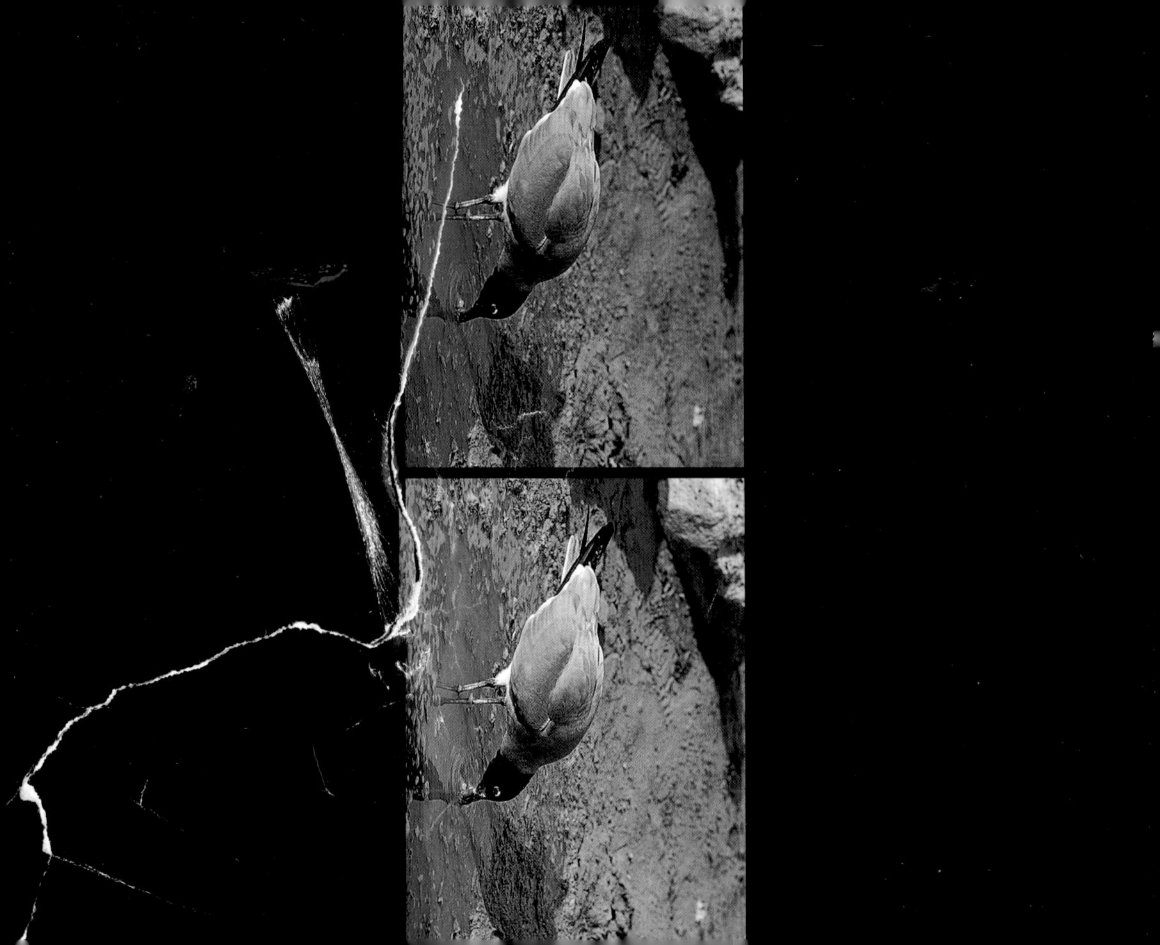

Plate 8

Marine Iguanas

Amblyrhynchus cristatus ssp. *Iguana Marina*

North Seymour Island

The marine iguana is the only seagoing lizard in the world. A vegetarian, it normally feeds at low tide on ulva, an algae growing in shallow reef and intertidal zones. This extraordinary lizard can dive to depths of 35 feet (10 meters) or more and remain underwater for up to an hour. Its tail is well adapted to swimming, while its sharp-clawed toes can grasp slippery rocks in surging seawater, and specialized teeth allow the marine iguana to graze from the rocky surfaces. Marine iguanas are known to spray excess salt out through their nostrils, a behavior that they sometimes use to warn away predators. The various islands are home to seven subspecies of marine iguana, which range from 2 to 3 feet (up to a meter) in length, tail included. During periodic El Niño weather patterns, when the iguana's green ulva algae food source gives way to toxic brown algae, its population is significantly reduced.

Plate 9

Red-Footed Boobies

Sula sula websteri *Piquero Patas Rojas*

Darwin Island

At around 2 pounds (1 kilogram), this endemic subspecies of red-footed boobies is the smallest of the three Galápagos booby species. Unlike the masked and blue-footed boobies, they have gripping feet that allow them to nest in bushes and trees. Here a flock of youngsters grip the railing of a ship, with Darwin's Arch visible in the background. As they mature, their plumage will lighten, their bills become blue, and their feet will turn red. Although the red-footed booby is the most abundant of the boobies (population 250,000), it is seen less often because its territory is limited to only the islands at the edges of the archipelago where its main predator, the Galápagos hawk, is absent.

Plate 10

Pacific Green Sea Turtle

Chelonia mydas agassisi *Tortuga Verde de Mar*

Wolf Island

The male green sea turtle is distinguished from the female by the shape of his shell and the length of his tail. The male is smaller than the female turtle and has a concave underside to his shell (plastron). His tail is considerably longer than the female's and also serves as the male's reproductive organ. Pacific green sea turtles are the only resident sea turtles in the Galápagos. While living and breeding in the archipelago, these mysterious reptiles also travel. Individual turtles have been found as far as 1,300 miles (2,100 kilometers) from where they were tagged. An endangered species, the Pacific green sea turtle is protected in the Galápagos and its population is doing well.

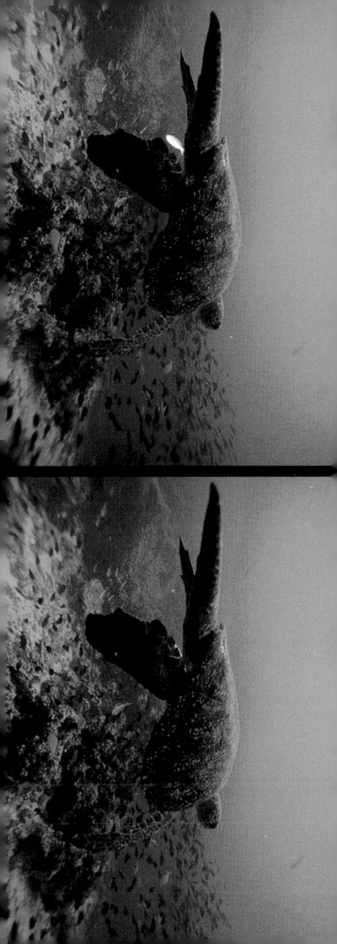

Plate II

Great Frigatebird

Fregata minor ridgwayi · *Fregata Común*

Isla Lobos

A resident seabird, the great frigatebird feeds far offshore, reducing competition with the closely related magnificent frigatebird *(Fregata magnificens)*. Its large wingspan is well adapted for gliding long distances. Although this spectacular bird snatches squid, fish, and crustaceans from the water while in flight, it lacks the oil glands necessary to waterproof it for actual seawater landings. In addition to fishing, the great frigatebird is a cleptoparasite, an animal that steals the food of others. It breeds in colonies and has a relatively long breeding cycle which lasts almost two years, and which begins with an elaborate courtship display. When the male great frigatebird successfully attracts a female, the courtship continues with the building of a crude nest platform, such as on this saltbush. One egg is laid and incubated by both parents for 55 days. Development of the chick is very slow, and the young bird is unable to fly for its first five months. Even then, a youngster like the one pictured will remain dependent on its parents for protection and food for more than a year while it learns the skills necessary for survival.

Plate 12

Galápagos Sea Lions

Zalophus californianus wollebacki *Lobos Marinos de Galápagos*

Isla Lobos

Many scientists consider the Galápagos sea lion to be a subspecies of *Zalophus californianus,* but distinguished from the California subspecies by its smaller size. Sea lions are able to survive in the equatorial climate of the Galápagos Islands due to the coldwater upwellings of the Humboldt Current. These nutrient-rich upwellings bring tolerable water temperatures and plentiful fish for the sea lions, who are especially fond of sardines. Locally, the sea lions are known as *lobos marinos,* or sea wolves. Isla Lobos, or Sea Lion Island, is named for the great number of sea lions that colonize its rocky shores. In this photo a female, or cow, relaxes with her pup.

Plate 13

Galápagos Sea Lion Pup

Zalophus californianus wollebacki *Lobo Marino de Galápagos, Cachorro*

North Seymour Island

Sea lions mate in the water, but the females give birth to pups on land. Mothers and pups, which weigh 11 to 13 pounds (5 to 6 kilograms), recognize each other by smell and by the distinct sound of their cries. After five weeks, the pup's first fur gives way to an adult pelt. Nursed on rich milk from its mother's teats, it will grow to four or five times its birth weight in a year. Mothers care for pups for one to two years at most, after which youngsters must fend for themselves. The sea lion's strong front flippers, which it uses for swimming, can also support it as it walks on the ground. Seals, in contrast, swim with their posterior flippers.

Plate 14

Galápagos Shark

Carcharhinus galapagensis *Tiburón de Galápagos*

Wolf Island

Galápagos sharks range the tropical oceans of the world, but are most common around oceanic islands such as the Galápagos. They are especially numerous at Wolf Island in the northern archipelago, where the sharks frequently swim in loose groups. Here they cruise the rocky submarine slopes, occasionally stopping for a cleaning by the barberfish *(Johnrandallia nigrirostris)*. The Galápagos shark is one of the five shark species known to display an aggressive posture, probably in response to a perceived danger, and sometimes as a prelude to an attack. Although there are about 205 shark species in the order *Carcharhiniformes,* a few species like the Galápagos shark have given the order the common name "requiem," meaning a Mass, for the dead. Galápagos sharks generally feed on fish, but they also eat sea lions and marine iguanas. These sharks are threatened by fishermen who exploit them for various products.

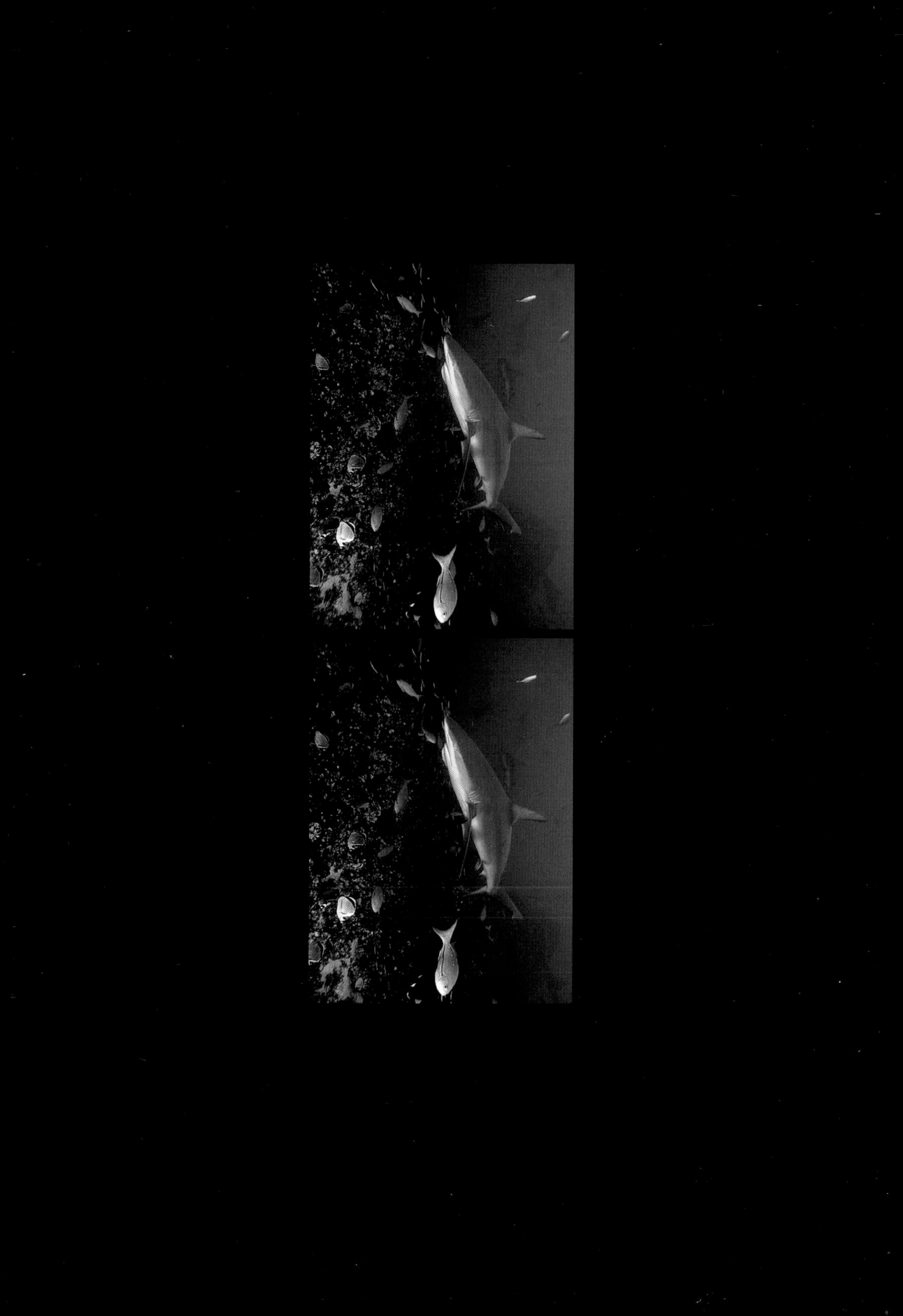

Plate 15

Spotted Eagle Rays

Aetobatus narinari *Chucho Pintado*

Cousin's Rock, Santiago Island

Four species of rays in the *Myliobatidide* family inhabit the Galápagos, known in Spanish as *rayas águilas*. All ray species have enlarged pectoral fins, also called wings, that give them a flattened, disklike shape. While the stingrays are bottom-dwellers, the eagle ray, cownose ray, and manta ray are known as "flying" rays because they swim in open water when not feeding. The eagle ray feeds mostly on bottom-dwelling mollusks but also eats other invertebrates such as sea urchins, which it crushes with its bony jaws. Swarms of fish follow rays to feed on the messy scraps as they eat. Eagle rays frequently swim in small to moderate-size schools, but usually feed alone. They can grow to 8 feet (2.5 meters) in width, although individuals over 4 feet (over 1 meter) are uncommon in the Galápagos. Their tails, with poisonous spines at the base, are more than three times the animals' body length.

Plate 16

Blue-Footed Boobies in Flight

Sula nebouxii excisa *Piquero Patas Azules*

North Seymour Island

These endemic seabirds are abundant throughout the archipelago. Their local name *piquero,* or lancer, derives from their long, pointed beaks. They feed by dropping into the water at high speed and scooping up fish on the way back to the surface. The blue-footed booby's long tail lets it turn sharply in the water after a dive, avoiding collision with the sea floor in the shallows. The lighter male bird is especially mobile under water, and so does more fishing than the female. All three species of booby are generally unafraid of humans and allow close approach. Despite its grace and agility in flight and under water, the booby has difficulty taking off unless it's perched on the edge of an outcropping or prominence that gives it plenty of room to become airborne.

Plate 17

Blue-Footed Booby with Chicks

Sula nebouxi excisa *Piquero Patas Azules*

North Seymour Island

Booby courtship takes place at a flat nesting site on the ground. The male brings twigs or feathers to the largely symbolic nest, and here the boobies perform a ritual song and dance in which the male whistles and the female honks. The female lays one to three eggs on the ground and incubates them with her large blue feet, which are generously supplied with blood vessels and help keep the eggs warm. If food is scarce, the firstborn chick (the most visible in this image) will usually get more and starve out its siblings, ensuring that at least one will survive. The nest area is surrounded by a ring of guano (bird excrement). If a chick leaves this area, it will not be allowed back in; probably as a means of preventing parents from raising another's young. The surviving chicks fledge after approximately a hundred days and are fed by their parents for another six weeks.

Plate 18

Landslide

Punta Vicente Roca, Isabela Island

The Galápagos Islands rise from the Galápagos Platform, a submarine plateau that slopes from 1,000 to 3,000 feet (360 to 900 meters) below the ocean surface. The surrounding ocean floor is as deep as 8,100 feet (3,200 meters). Islands such as the Galápagos that rise from the deep sea floor are known as oceanic islands. Volcanic activity and plate tectonics have built the Galápagos Island chain over the last ten to fifteen million years. New islands continue to emerge as older islands migrate eastward with the Nazca Plate on a collision course with South America. A new island is presently, but slowly, rising from the sea floor west of Fernandina. Isabela Island, seen here, is composed of no fewer than six volcanos. This photograph captures a landslide eroding the tuff cliffs of one of them, Volcan Ecuador.

Plate 19

Hornshark

Heterodontus quoyi · *Gato*

Banks Bay, Punta Vicente Roca, Isabela Island

The common name of this strange-looking shark refers to its hornlike dorsal fin spines, which appear here as white spots. These small creatures, about 3 feet (1 meter) long, have been referred to by scientists as "living fossils" because their unique teeth resemble certain ancient species known only from fossils. These sharp front teeth capture swimming prey, and flat rear teeth crush mollusks and crustaceans gathered in by the fatty appendages around the mouth. The distinctive egg cases of hornsharks are screw-shaped with broad, spiral flanges. The female shark carries the egg case to a rock crevice and wedges it where it has a chance to harden. A juvenile shark emerges nine to twelve months later. Hornsharks, which are most common in the western archipelago, move about slowly on rocky and sandy bottom areas to a depth of 65 feet (20 meters). Whether the Galápagos population is native depends on the unresolved question of whether it is distinct from the species *Heterodontus peruanus,* found along the mainland of South America.

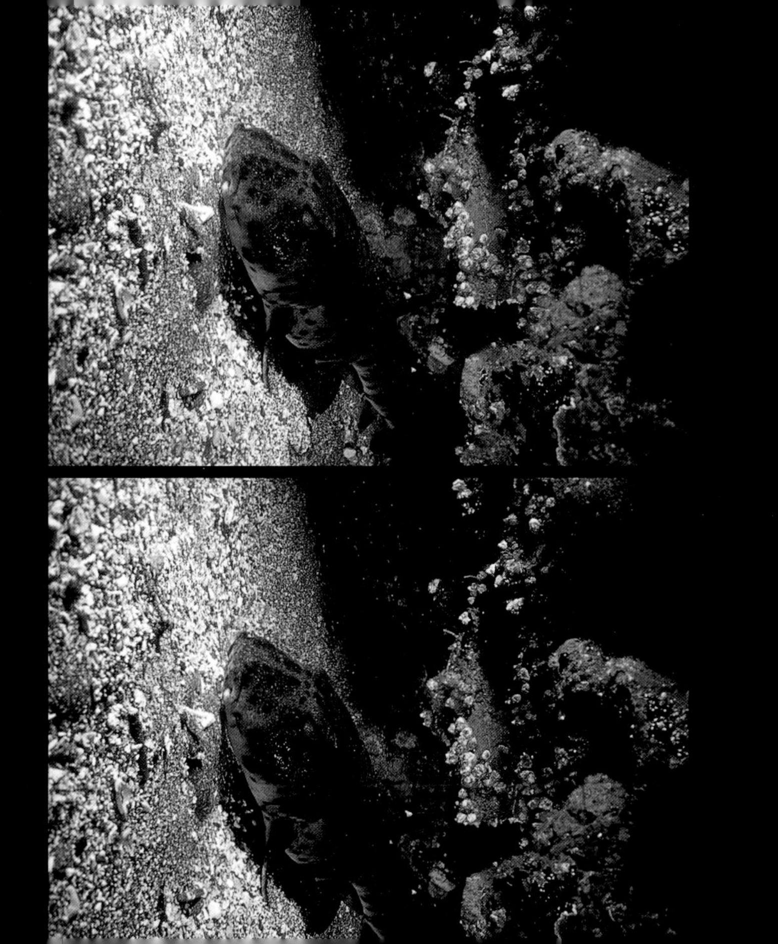

Plate 20

Sea Hare

Aplysia cedrosensis *Conejo de Mar*

Banks Bay, Punta Vicente Roca, Isabela Island

The sea hare, a mollusk lacking the protection of a shell, has developed the ability to defend itself by the secretion of inks. The common name "sea hare" arises from the two appendages *(rhinophores)* that resemble rabbit ears behind the head. In this close-up you can also see the small eyes. The two fleshy dorsal flaps *(parapodia)* on its back are extensions of its foot, and enclosed within them are its gills. Sea hares are hermaphroditic, meaning that each animal has male and female reproductive organs. They are not self-fertilizing, however, and individuals often cluster together in "daisy-chains" to mate. Afterwards, a single sea hare can deposit up to 180 million eggs. The sea hares' large nerve cells make them extremely important to medical researchers interested in studying nerve responses.

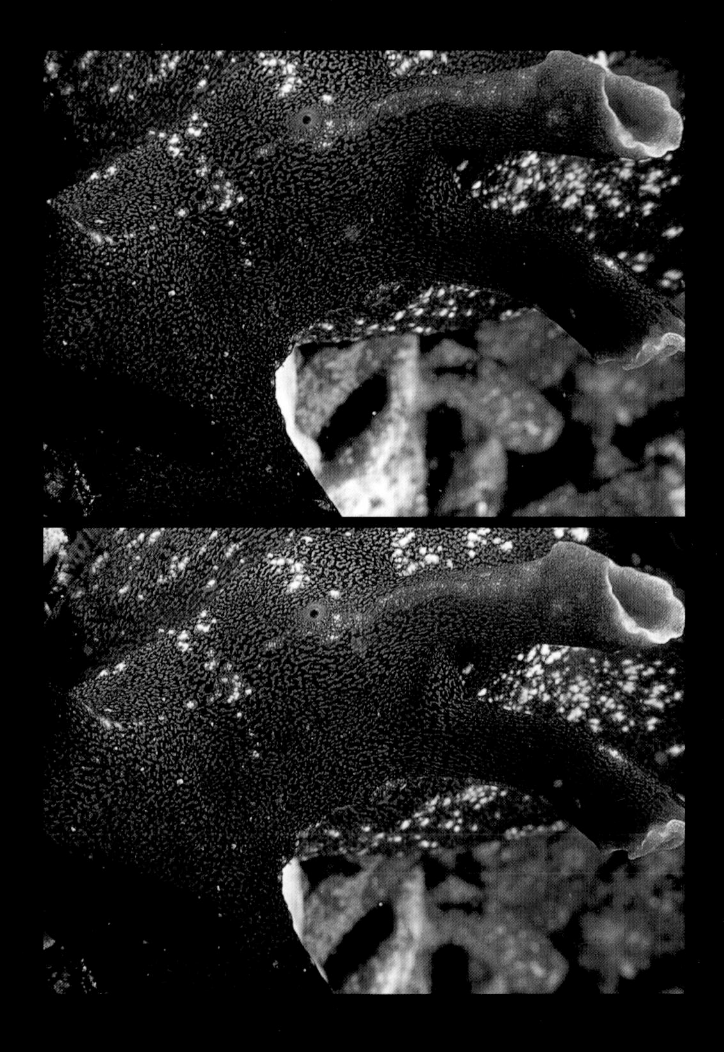

Plate 21

Hingebeak Shrimp

Rhynchocinetes typus *Camarón*

Banks Bay, Punta Vicente Roca, Isabela Island

The hingebeak shrimp is one of approximately two thousand shrimp species inhabiting the world's oceans. Shrimp, including the hingebeaks, have especially well-developed vision, with eyes located on movable stalks. Hingebeaks are reclusive creatures that hide during the day on the rocky bottom, a response to the pressures of daytime hunting by fish and other predators. Also known as "night shrimp," they are most active after sundown, as their nickname suggests, when they emerge from their retreats to feed in relatively safer surroundings.

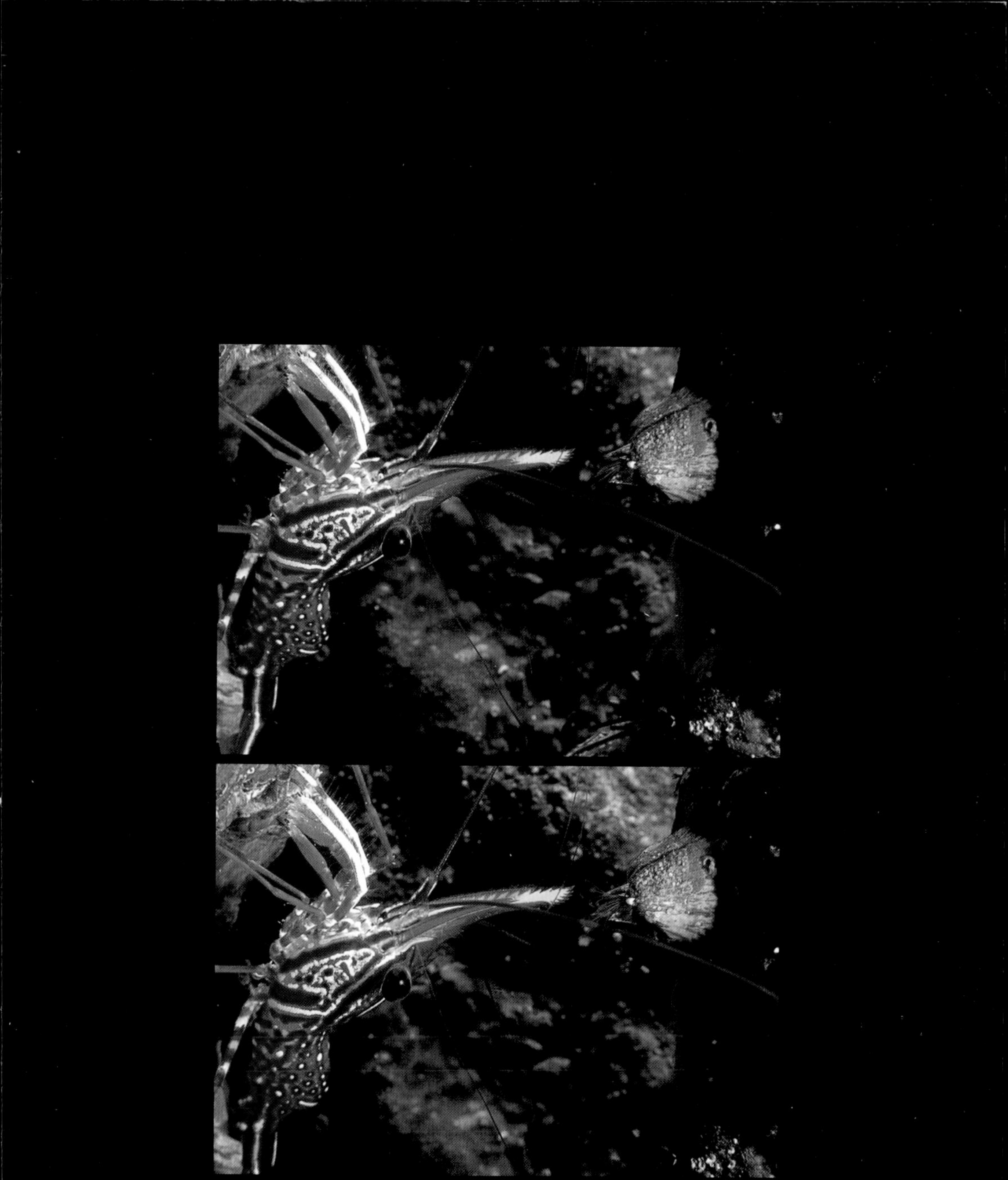

Plate 22

Brown Pelican

Pelicanus occidentalis *Pelícano*

Punta Vicente Roca, Isabela Island

One of only two pelican species found in the Western Hemisphere, the remarkable brown pelican has changed little in the last thirty million years. Despite this testament to its endurance, the brown pelican's survival has been severely challenged in the last century, and it is now designated an endangered species. Among the world's largest birds, the brown pelican must struggle to take off. Once airborne, however, these birds glide and soar effortlessly on thermal currents or ride the valleys between ocean waves. They feed in flight by diving onto schools of small fish. At the precise moment that the bird's bill hits the water, the pelican opens its pouch and scoops up the target fish. Linked to the sea, brown pelicans do not migrate, but remain close to their oceanside colonies all year round, where nesting and courtship take place.

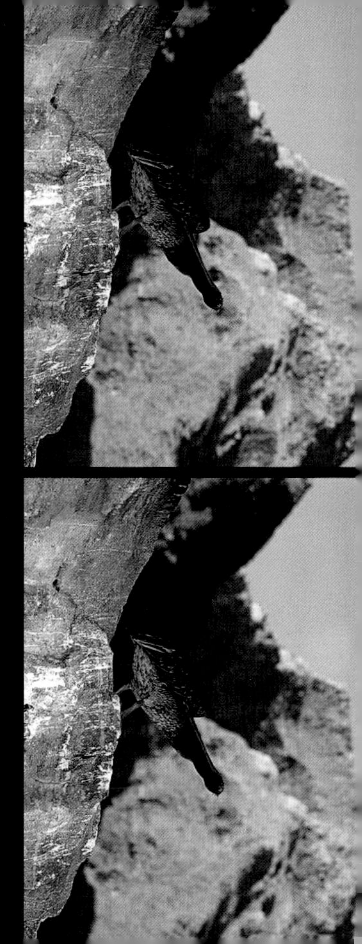

Plate 23

Volcanic Landscape

Bartolomé Island

Bartolomé is the smallest of the islands in the western archipelago, only three-quarters of a square mile (1.2 square kilometers) in size. The island is a landscape of volcanic desolation. While the islands in this western group are the youngest, about two million years old, the stark volcanic landscape of Bartolomé suggests that it is much younger. Only the lava cactus and a few forms of vegetation grow presently on the slopes of the volcano. In this view to the west, we see the projection of Pinnacle Rock, and behind it Sullivan Bay and Santiago Island. A small colony of Galápagos penguins makes its home in the coves on either side of Pinnacle Rock.

Plate 24

Galápagos Penguins

Spheniscus mendiculus ❋ *Pingüinos de Galápagos*

Pinnacle Rock, Bartolomé Island

This small endemic seabird, related to the Magellan penguin *(Spheniscus magellanicus),* is the only penguin to live and breed as far north as the equator. Its ancestors arrived in the Galápagos via the cold Humboldt Current from Chile. As in other species in the order *Sphenisciformes,* the wings of the Galápagos penguin have evolved for swimming, not for flight. The bird's population of about 13,000 was devastated by El Niño in 1982–1983 and is slowly recovering. These shy animals are found in small groups of two to seven birds concentrated along the coast of numerous islands. In their six-month breeding period, a pair of penguins can produce up to three clutches of eggs, depending on food availability. Each clutch normally consists of two eggs laid in a rock cavity. Both parents share in the incubation of the eggs over about 40 days. Usually only one chick fledges, and the parents care for it for about two months.

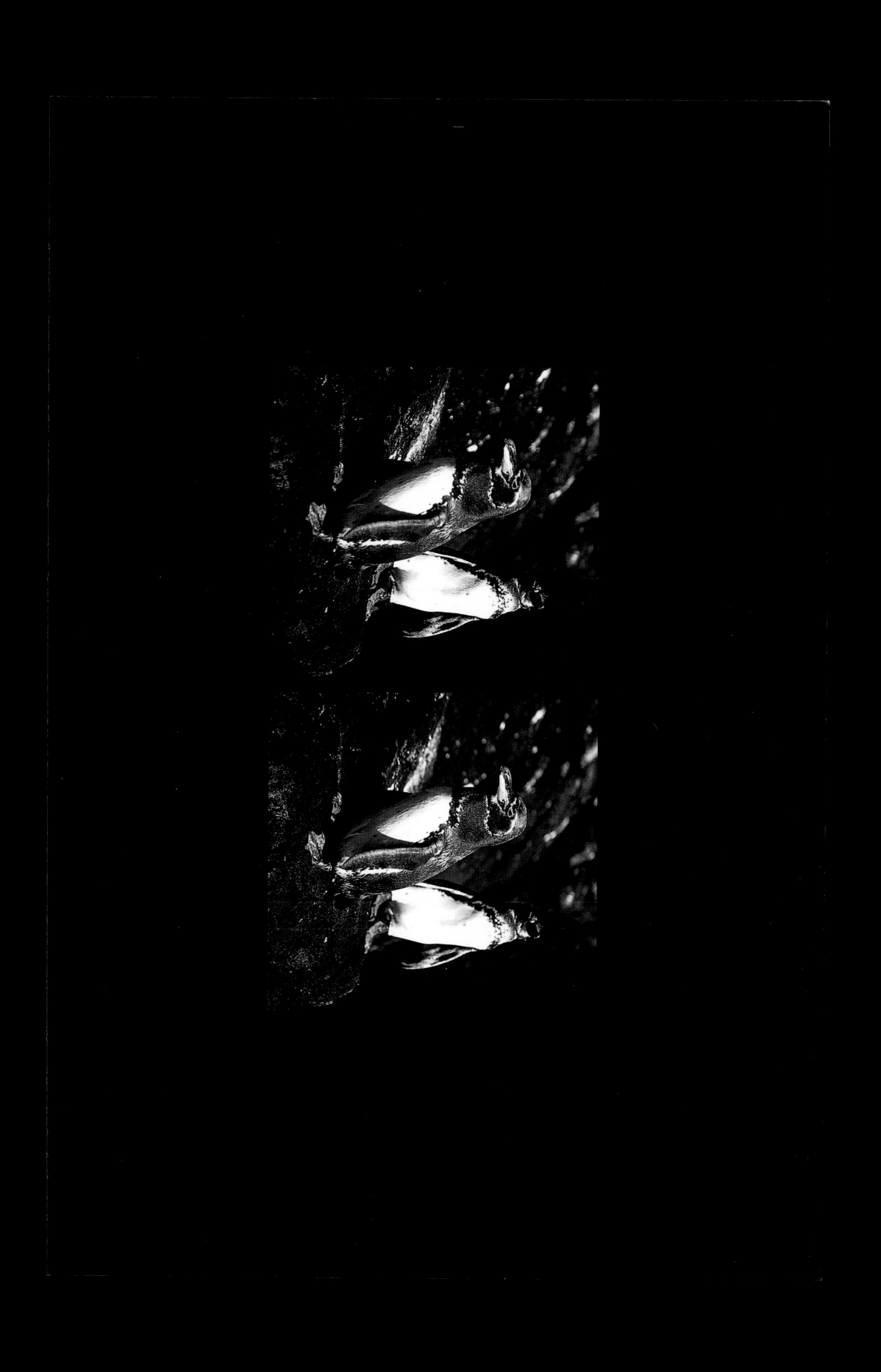

Plate 25

Sally Lightfoot Crabs

Grapsus grapsus *Zayapas*

Bartolomé Island

This common crab was named the Sally Lightfoot crab by sailors in honor of a nineteenth-century dancer. These highly visible creatures are colored in riotous hues of bright red and yellow, with blue undersides. Sally Lightfoot crabs, which live on the shore but do not appear to venture far under water, are climbing crabs that cling to the wave-swept rocks with strong, spiny legs. *Grapsus grapsus* eats various foods, including seaweed and marine iguana carcasses, helping to clean the shoreline. They are, in turn, hunted by birds. These comical crabs crawl blithely over live iguanas basking on the rocks in the hot equatorial sun, while the lizards do their best to ignore them, perhaps because the crabs sometimes remove old skin from their iguana hosts.

Plate 26

Fur Seal

Arctocephalus galapagoensis *Lobo de Dos Pelos*

Cousin's Rock, Santiago Island

Despite their name, these animals are sea lions, not seals. Known locally as *lobo de dos pelos* (meaning "double-fur sea wolf"), this subspecies of sea lion is endemic to the Galápagos Islands. Fur sea lions reach about a third of the weight of the larger Galápagos sea lions. They were hunted to near extinction for their fur in the early part of the twentieth century; today their protected population has climbed to more than 30,000 individuals. The fur seal is the only member of the genus *Arctocephalus* to be found in the tropics. Due to their extremely warm coat of fur, they are less tolerant of heat than Galápagos sea lions. Fur seals prefer cooler waters and steep rocky shorelines, where they may shelter in the shade. The large eyes of the fur seals help them to see at night when they leave the colony to hunt squid and fish.

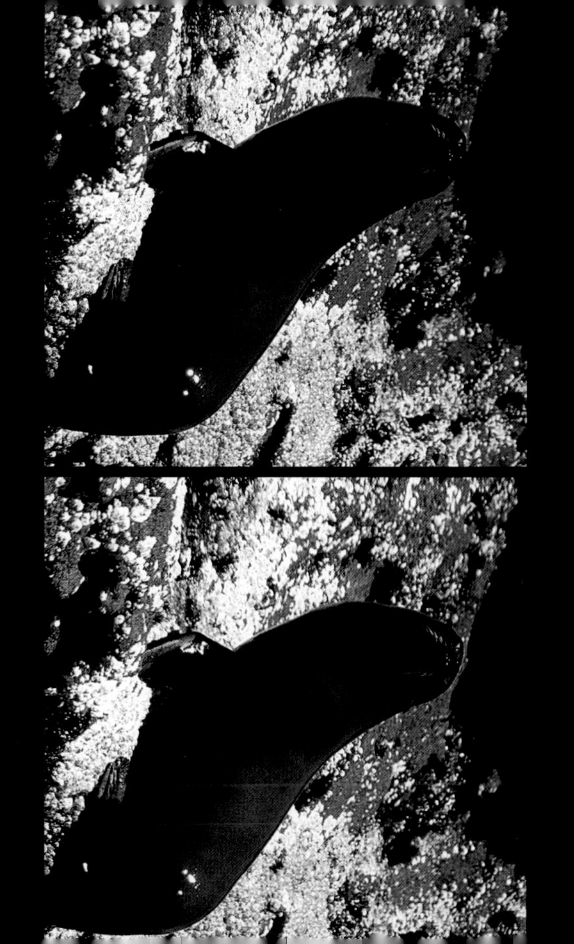

Plate 27

Fine Spotted Moray

Gymnothorax dovii ❋ *Morena Puntofino*

Wolf Island

This member of the large moray eel family is common throughout the archipelago, inhabiting rocky areas and walls. But unlike many other species whose heads can be seen protruding from crevices, the fine spotted moray often displays ostrich-like behavior, putting its head in a hole and leaving the rest of its body exposed on the reef. The gaping mouth in this image is not intended as a menacing gesture; the animal must constantly open and close its mouth to pump water over its gills for respiration. Strong jaws, fang-like teeth, and leathery skin make morays highly effective predators. At night they emerge from their rocky lairs to hunt in the open.

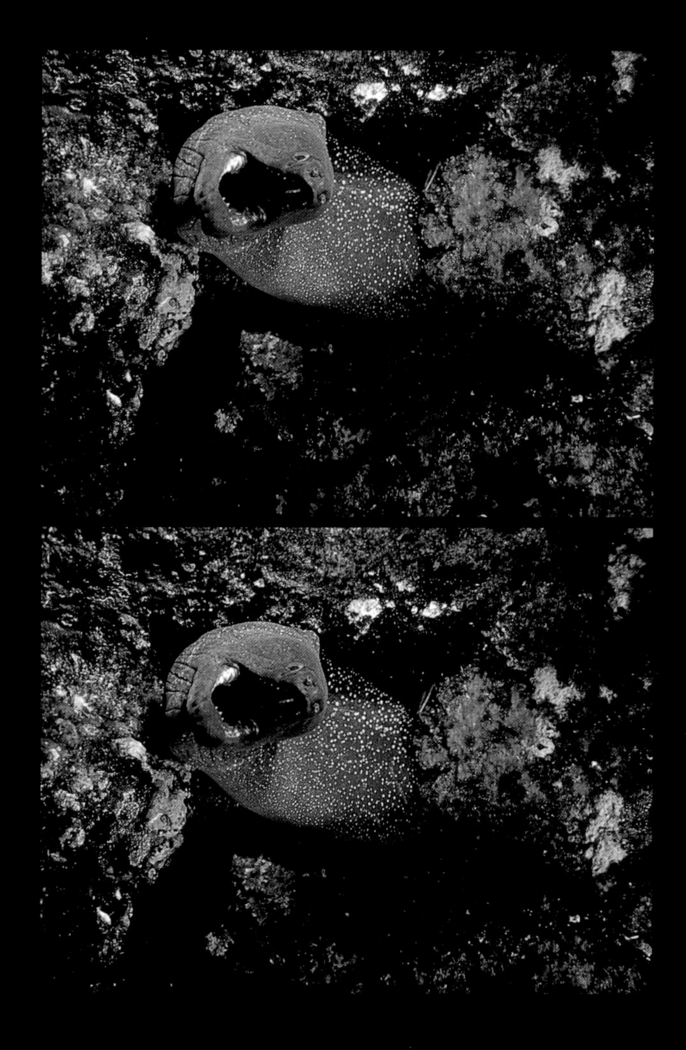

Plate 28

Barracuda

Sphyraena idiastes *Barracuda*

Cousin's Rock, Santiago Island

In this photo a school of barracuda holds position facing into the swift current that sweeps around the west end of Cousin's Rock, a tiny speck off the northeast coast of Santiago Island. By approaching slowly, a diver may enter the school. These fearsome predators feed mainly on squid and other fish, hunting mostly at dawn and dusk. All 20 species worldwide are similar in appearance, characterized by a long, streamlined, silvery body with long jaws and protruding teeth. *Idiastes* is the only species found in the Galápagos Islands. It is a relatively small barracuda, averaging 1 to 2 feet, with a maximum of 3 feet (1 meter). As with some other barracuda species, *idiastes* may cause chevron bar markings to appear along its side at will. The barracuda lays small (1-millimeter), round eggs. The larvae live in the deep ocean, and the young fish settle in sheltered coastal water.

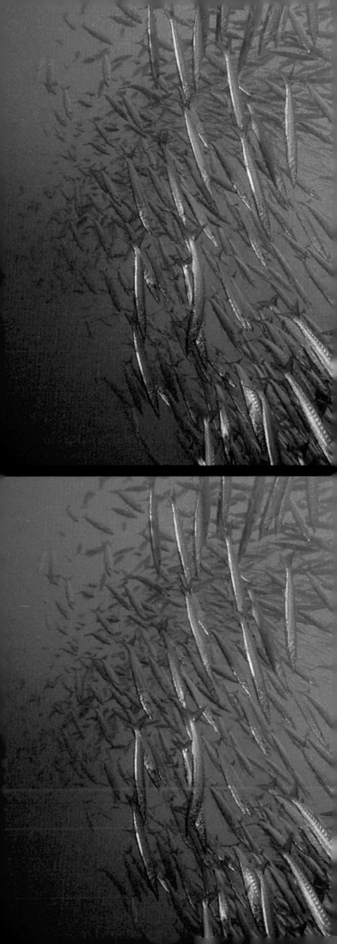

Plate 29

Vegetative Zones

Santa Cruz Island

From a vantage point in the area known as the brown zone of Santa Cruz Island, this view takes in five vegetative zones. The brown mosses are liverworts, which hang from the Scalesia trees. They are epiphytes, plants that derive their moisture and nutrients from air. The common name "brown zone" refers to the color of these mosses. This view looks out over the Scalesia zone, the island's main agricultural region on the island, and beyond it the transition zone, the arid zone, and the littoral zone. The tall flattop trees in the distance are non-native balsa trees.

Plate 30

Giant Tortoise

Geochelone elephantopus *Galápago*

Santa Cruz Island

Male Galápagos tortoises, much larger than their female counterparts, reside most of the year around the rims and in the calderas of volcanoes. For much of the year, they feed contentedly on about fifty species of vegetation. With the onset of the wet season, males engage in ritual fighting before descending the slopes in search of mates. At the lower elevations, tortoises of both sexes may be found near ponds, such as this dome-shaped specimen in the Scalesia zone of Santa Cruz Island. After mating, the females must slowly travel all the way down to the coastal arid zone to lay their eggs. Sometime during the six-month period after fertilization, the female turtle excavates a nest and deposits a single batch of eggs. This egg clutch is sealed with bodily secretions and capped with soil. The temperature in the nest determines the sex of the offspring. If it reaches 84 degrees Fahrenheit (29 degrees centigrade) or higher, the eggs will hatch female tortoises, while lower temperatures will produce males.

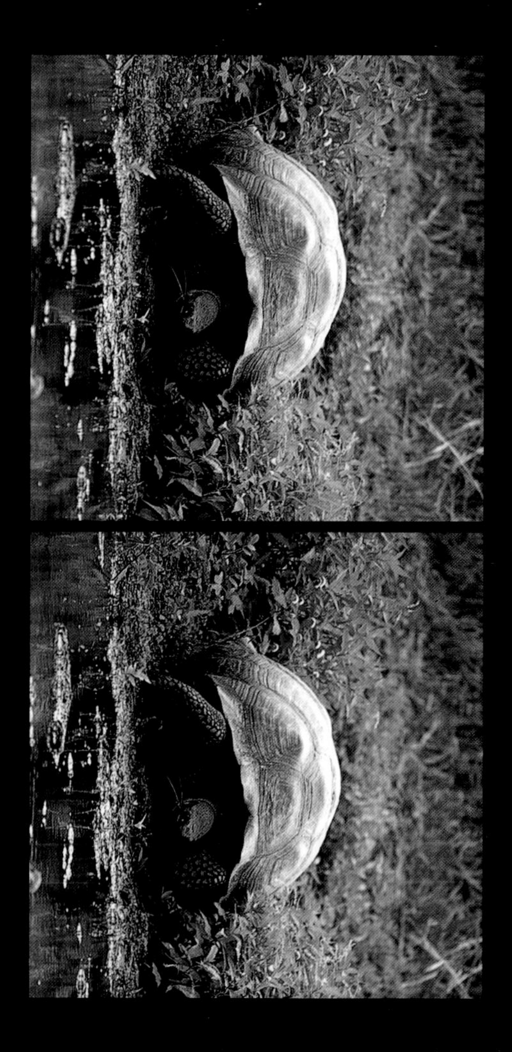

Plate 31

Giant Tortoise

Geochelone elephantopus *Galápago*

Charles Darwin Research Station, Santa Cruz Island

These peaceful giants, which can weigh over 500 pounds (227 kilograms), are as mysterious as they are huge. No one has been able to explain how this land animal made its way to the Galápagos Islands, which have never been connected to any continental land mass. Hundreds of thousands of tortoises were found 400 years ago when pirates arrived and started to use them for food (the tortoises survived for as long as a year on pirate boats due to their ability to store water and fat). Ten subspecies survive in viable populations today. The sole survivor of an eleventh subspecies, named Lonesome George, lives at the Charles Darwin Research Station. The greatest threat to the tortoises comes from non-native predators introduced to the islands such as rats, pigs, and goats. The capture and breeding program at the Charles Darwin Research Station has helped to perpetuate the survival of these astonishing creatures. Pictured above is a "saddleback" tortoise, a variety that lives on the drier, lowland islands where its long neck helps it reach higher for the scarce vegetation.

Plate 32

Opuntia Cactus

Opuntia echios

South Plaza Island

The opuntia, or prickly pear cactus, seen in this photograph from South Plaza Island, is one of three Galápagos cactus species that comprise the dominant vegetation in the arid zone, one of seven distinct vegetative zones in the Galápagos Islands (coastal or littoral, arid, transition, Scalesia, brown, Miconia, and pampa or fern zone). Only the biggest islands, such as Santa Cruz, have all seven vegetative zones. The arid zone may extend from a few feet from shore upwards to an elevation of roughly 260 to 400 feet (80 to 120 meters), but may reach higher on the dry leeward side of the islands, to over 1,000 feet (300 meters). Beneath the cacti in this photograph are beds of Sesuvium flowers. The opuntia cactus, which can be either shrub- or treelike, can reach 16 feet (5 meters) in height and is armed with sharp spines at the base to defend itself against land iguanas and other animals that might eat it.

Land Iguana

Conolophus subcristatus *Iguana Terrestre*

South Plaza Island

This lizard is one of two species of land iguana found only in the Galápagos Islands. An impressive 3 feet (1 meter) long, these iguanas are mainly vegetarian, favoring opuntia cactus fruit but also feeding on insects, carrion, and sea lion afterbirth. Each isolated population of land iguanas has developed its own unique breeding season. Males are highly territorial and will breed with up to seven females. After breeding, the females fight to establish burrows as nesting sites, where they deposit six to fourteen soft-shelled eggs. The female fills in the burrow with soil and guards it for a week. If the young iguanas survive predation by hawks and snakes, they may grow to as much as 25 pounds (11 kilograms) and live an amazing 60 to 70 years. On Fernandina Island, the females nest in soft ash inside the caldera. Occasional volcanic activity destroys these sites with a disastrous effect on the population. A breeding program begun in 1976 at the Charles Darwin Research Station on Santa Cruz Island has helped to preserve the species.

Plate 34

Stone Scorpionfish

Scorpaena plumieri mystes *Brujo*

Banks Bay, Punta Vicente Roca, Isabela Island

This young animal, photographed at very close range, displays the typical characteristics of the scorpionfish in exquisite detail. Its large, stout head is mottled in color with numerous fleshy flaps and flexible appendages called *cirri*, all of which aid in camouflage among the rock outcroppings. The stone scorpionfish can also darken or change color at will to avoid detection. Nearly invisible when it wants to be, the scorpionfish remains still on the bottom and moves only when disturbed. Although they are unaggressive towards humans, scorpionfish possess venomous dorsal spines that can inflict a painful sting, and their excellent camouflage makes them a real danger for divers who don't watch where they step.

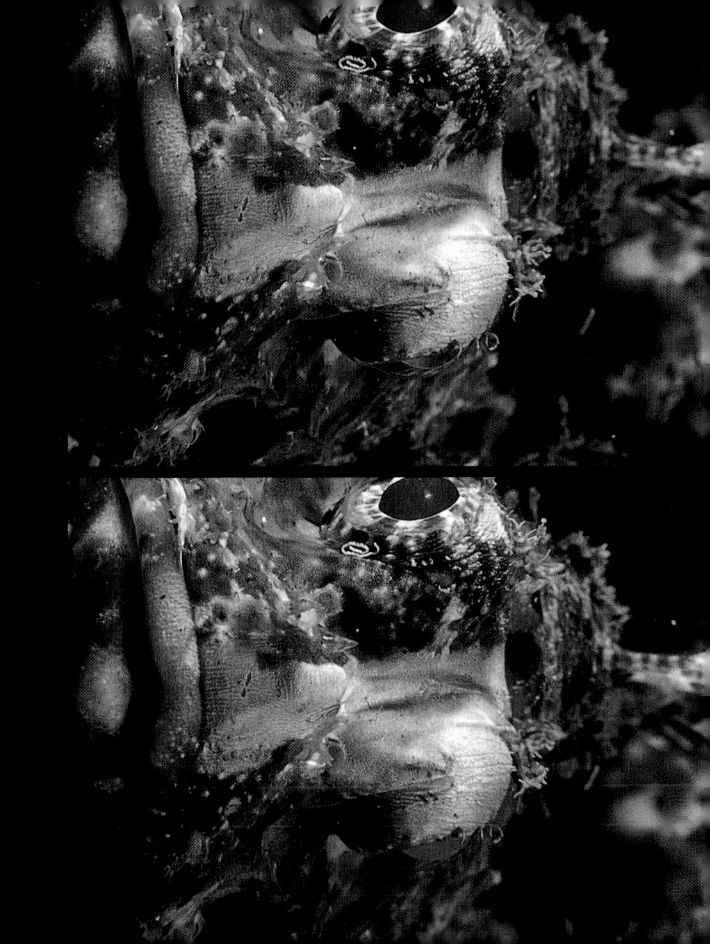

Plate 35

Pacific Seahorse

Hippocampus ingens *Caballito de Mar del Pacifíco*

Cousin's Rock, Santiago Island

The Pacific seahorse is found only in the eastern Pacific Ocean and is the only seahorse species in the Galápagos Islands. It is seldom seen in the islands. This one was found about 50 to 60 feet (15 to 18 meters) below the surface on the south side of Cousin's Rock, its tail curled around the branches of a black coral tree, which is the yellow substance in the background. Seahorses often tuck their heads into the coral branches of their resting spots, making them difficult to photograph. When not occupying coral and reef habitats, the Pacific seahorse floats in the open ocean. The large Pacific seahorse can grow to nearly 10 inches (25 centimeters) in length and has a wide range of coloration. It is identified by the numerous fine line markings radiating from the eye and running along the body.

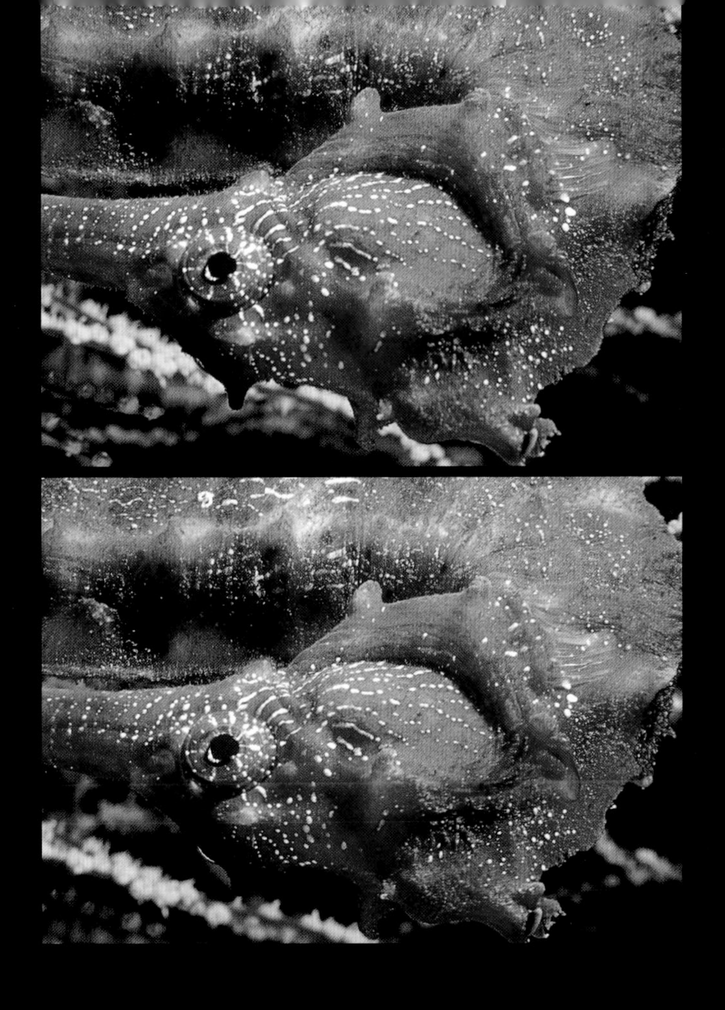

Plate 36

Sanguine Frogfish

Antennatus sanguineus *Pez Sapo Sanguíneo*

Cousin's Rock, Santiago Island

The common name "frogfish" arises from this bottom-dweller's ability to hop and crawl over reefs on leglike pectoral and pelvic fins. The coloring of these bizarre-looking fish is highly variable. With a chameleonlike ability to change color and pattern at will they are able to perfectly match their habitat. This one is trying to blend into a rocky recess of a ledge at a depth of 65 feet (20 meters) on the south side of Cousin's Rock. Frogfish are also commonly referred to as anglerfish due to their method of attracting prey with an elaborate fishing mechanism, as seen in the foreground of this photo. The first dorsal fin spine is a "fishing pole," called an *illicium,* topped with a fleshy "lure," or *esca*. When a victim has been lured in close by the twitching "bait," the anglerfish rapidly expands its mouth cavity up to 10 times its normal size, vacuuming up and swallowing fish as large as itself.

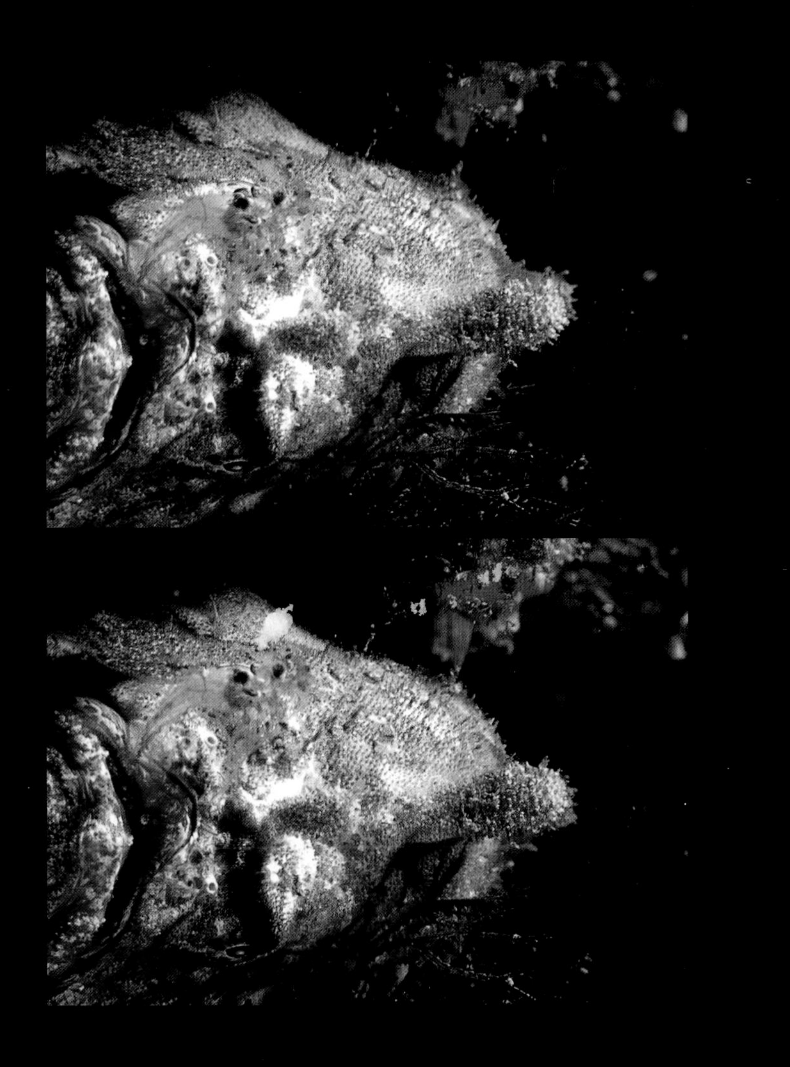

Plate 37

Miconia Zone

Santa Cruz Island

This vegetative zone is named after the native Miconia bush, *Miconia robinsoniana*. The bush (seen in the lower left-hand corner) is also known as *cacaotillo*, because the leaves look similar to those of the cacao plant. Miconia is found only on San Cristóbal and Santa Cruz Islands, where it occupies a zone just above the upper reaches of the Scalesia forests and just below the pampa zone. The Miconia zone lies within the humid influence of the special precipitation known as *garua*, and visibility can be extremely limited in the thick mists. On Santa Cruz Island a trail leads through the Miconia to Media Luna, the collapsed remnants of a volcanic cone covered in vegetation. On the slopes of Media Luna, Miconia bushes fight with Cinchona shrubs for vegetative dominance.

Plate 38

Scalesia Zone

Santa Cruz Island

This pond lies in the humid Scalesia zone on Santa Cruz Island. In this area during the otherwise dry season, from May to December, hot air encounters land and precipitates in a fine rain known as *garua*. This region also experiences a normal wet season, from December to May, when conventional rains occur. Such year-round precipitation supports an abundance of moisture-loving vegetation. By far the most common tree is the native *Scalesia pedunculate,* seen here in the background. Known locally as *lechoso,* it is an evergreen member of the sunflower family and grows in dense forests covered with moss and liverworts. In the pond is a giant female tortoise that has migrated up from the arid zone for the mating season.

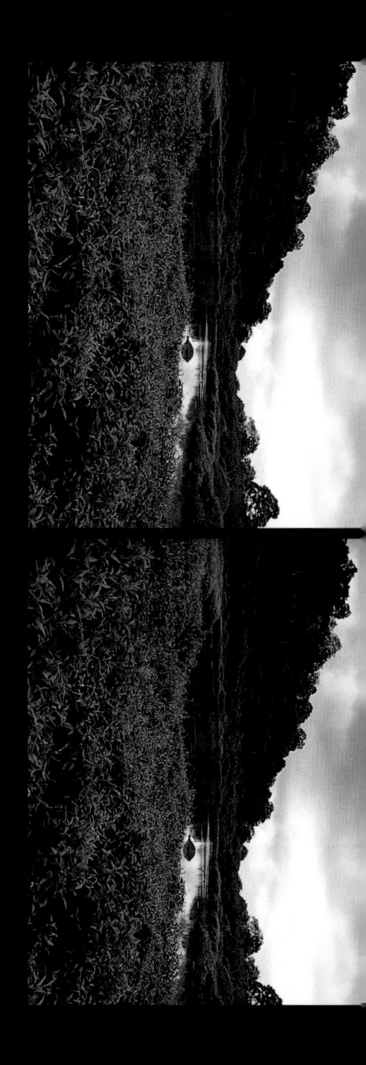

Plate 39

Ocean Cliffs

Wolf Island

Also known as Wenman Island, Wolf was the last island visited by the 26-year-old Charles Darwin on his first and only scientific expedition to the Galápagos, in 1835. Darwin, too, must have marveled at the sheer cliffs rising 740 feet (225 meters) above sea level. They are the remains of a caldera, the collapsed crater of a shield volcano. The tall spire, known as the chimney, is a lava plug left behind when the surrounding tube eroded away. These remote cliffs are a refuge for gulls, frigatebirds, boobies, and tropical birds, which perch on precipitous ledges. Wolf Island is also home to fur seals and a distinct species of "vampire finch," a bird which uses its sharp bill to peck through the skin of red-footed boobies and drink their blood as an adaptation to the arid climate. Beneath the sea surrounding this remote island numerous deep-sea predators gather, attracted to the strong currents that sweep the submerged volcanic slopes.

Plate 40

Pacific Bottlenose Dolphin Pod

Tursiops truncatus *Delfín*

Wolf Island

The Pacific bottlenose dolphin is the cetacean most commonly observed in the Galápagos Islands. Other species rarely come near people, but the curious and playful bottlenose dolphin enjoys riding the bow wave of boats. Observers are occasionally rewarded with a spectacular leap into the air. This image was captured as a pod of inquisitive dolphins led and flanked a small inflatable boat moving beneath the steep cliffs of Wolf Island. Some playful members of the pod swam close alongside the boat and slapped the water to send a salty spray over the passengers.

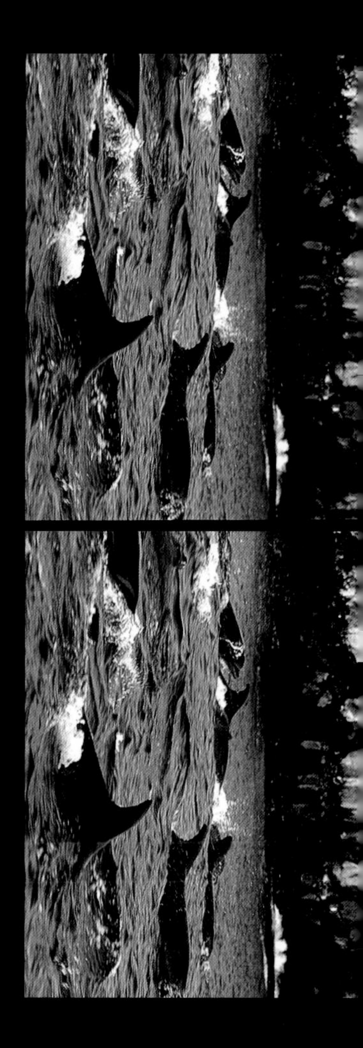

Plate 41

Pacific Green Sea Turtle

Chelonia mydas agassisi *Tortuga Verde de Mar*

Wolf Island

Eight species of marine turtle exist today. They are remarkably similar to their reptilian ancestors of 150 million years ago, a resemblance which suggests that they have adapted extremely well to life in the ocean. Male green sea turtles live out their lives entirely at sea, but females come ashore to lay eggs on the beach, where they dig a hole to deposit 50 to 110 eggs. When the eggs hatch, about two months later, the young turtles instinctively dash for the sea. Juvenile turtles are vulnerable to numerous predators on land and in the water, and only a very few survive into adulthood. Those that reach maturity have few enemies except for humans. Regrettably, human activity increasingly threatens the survival of these ancient mariners.

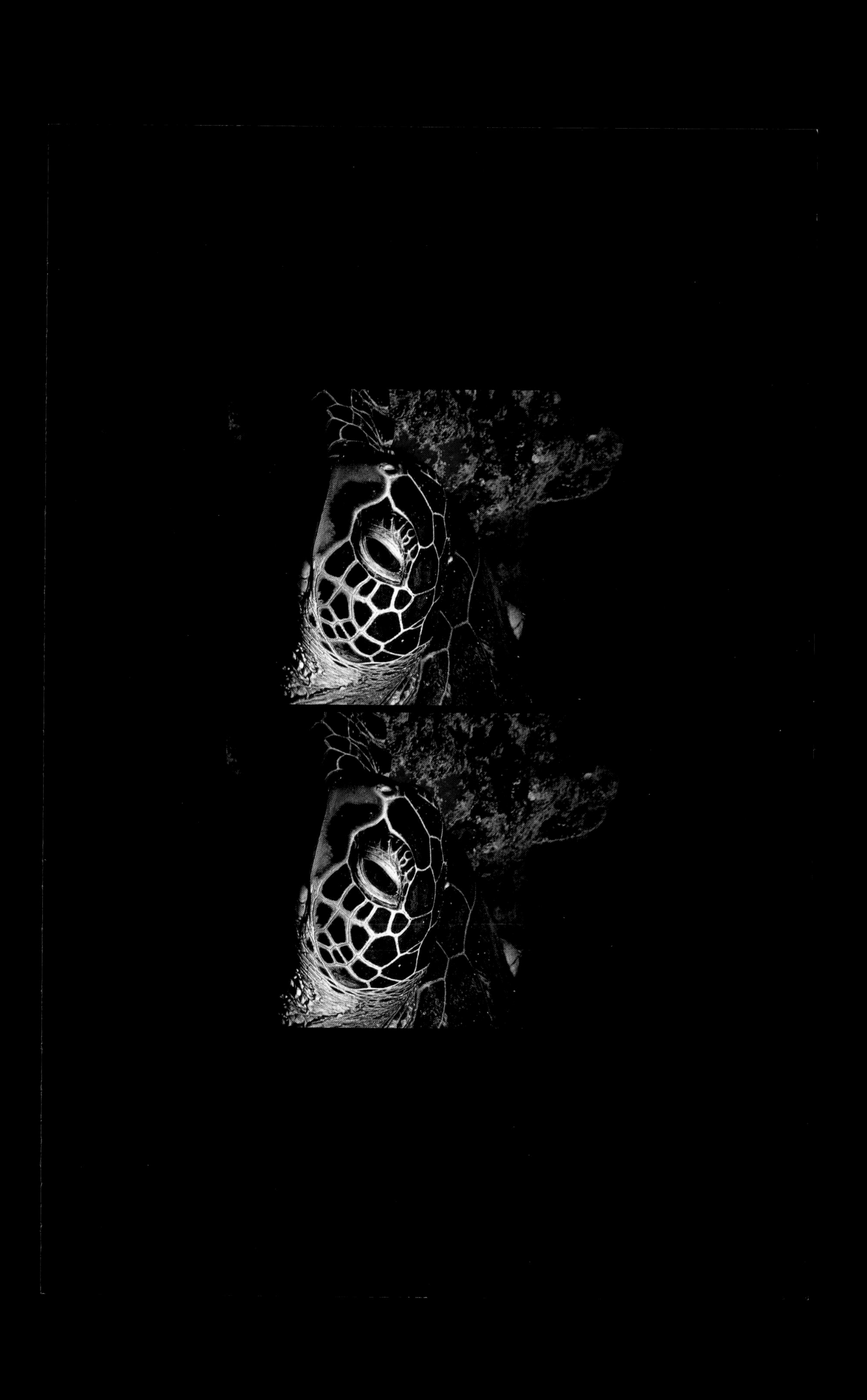

Plate 42

Almaco Jack

Seriola rivoliana *Palometa*

Wolf Island

Fish in the *Carangidae* family, known as jacks, are strong-swimming, robust fish that may grow to as much as 5 feet (1½ meters) in length. Usually silvery in appearance, almaco jacks are roving ocean predators, feeding on other fishes. In the Galápagos Islands, they gather in schools over outer reef banks and on slopes to depths of 180 feet (55 meters). Young jacks frequently shelter beneath floating debris in the water. Adults develop a dark band that runs from the dorsal fin across the eye to the lower lip. The almaco jack can also be seen swimming closely with larger animals such as whale sharks. This may be a camouflage technique which allows the jack to surprise and eat other fish.

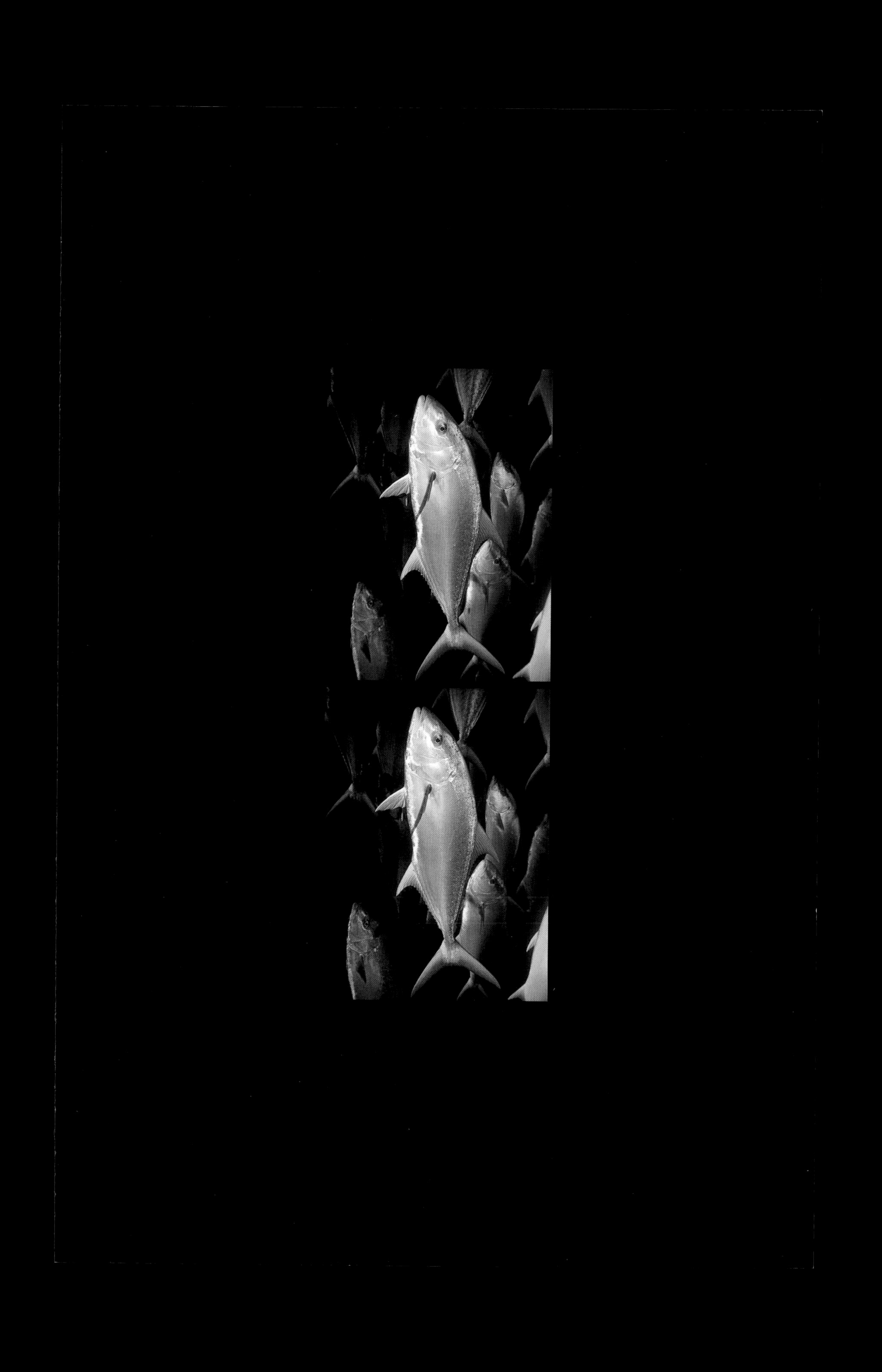

Plate 43

Marbled Ray

Taeniura meyeni *Raya Manchada*

Wolf Island

Marbled rays are quite common on the rocky slopes and sandy bottom off Wolf and Darwin Islands at depths of 15 to 90 feet (5 to 30 meters). Like most rays and skates, the marbled rays are flattened and swim with a graceful undulation of their pectoral fins. They can bury themselves in the sand for camouflage and draw in water from a small hole behind each eye, expelling it through gills on their undersides. Marbled rays feed at night on fishes, mollusks, and crustaceans. The rays at Wolf Island are large, often measuring more than 6 feet (2 meters) wide. Their tails are armed with a sharp, venomous spike. Their only major predators are sharks, who devour them with seeming disregard for the rays' poisonous weaponry.

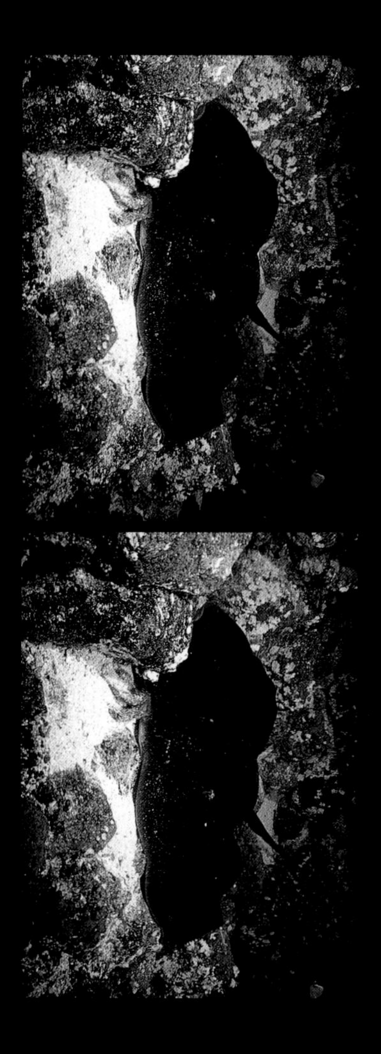

Plate 44

Scalloped Hammerhead Shark with Creolefish

Sphyrna lewini, Paranthias colonus *Tiburón Martillo, Gringo*

Wolf Island

The scalloped hammerhead is one of the three large species out of a total of eight hammerhead types comprising the genus *Sphyrna*. They can reach more than 13 feet (4 meters) in length and are quite common at the northern islands of Wolf and Darwin, where they swim in schools or small groups during the day. Although normally retiring, in the Galápagos Islands scalloped hammerheads seem to have no fear of divers and are often inquisitive. They feed on fishes, crustaceans, and other sharks, but are particularly fond of rays and skates, which they locate electromagnetically by sweeping their broad heads back and forth over the bottom in a method similar to using a metal detector. Scalloped hammerhead sharks do not seem to be adversely affected by the venomous barbs of stingrays. In this photograph, adult creolefish are seen aggregated in their typical daytime feeding pattern when they are seemingly unafraid of sharks and divers alike.

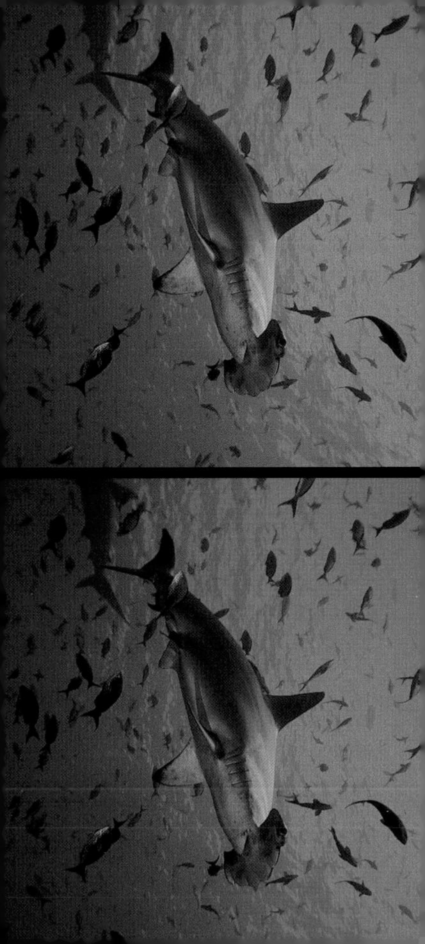

glossary

An **aggregation** is the collection of many individuals into a group.

Appendages are subordinate body parts, like the limbs, of an animal.

An **archipelago** is a group of islands.

An **arid zone** is an excessively dry area characterized by scant rainfall.

Bottom-dwellers are organisms that live predominately on the ocean floor.

A **caldera** is a collapsed shield volcano.

Camouflage is the use of color and pattern to help an animal or plant blend into its surroundings.

Cetaceans are aquatic mammals, like whales, dolphins, and porpoises, that have torpedo-shaped, nearly hairless bodies; paddle-shaped forelimbs; no hind limbs; a nare, or blowhole, at the top of the head; and flattened tails.

Class is a category used to classify animals. Order, family, genus, and species are other categories of biological classification.

Dorsal corresponds to the location near or on the back of an animal.

El Niño is an irregularly occurring disruption of the ocean-atmosphere system in the tropical Pacific affecting weather around the globe.

Extinction occurs when no members of a species are left alive.

To **fledge** is to grow the feathers necessary for a bird to fly.

Garua is a special, humid precipitation that causes thick mists to form at certain elevations.

Genus is a category of biological classification, ranking below the family and above the species.

Gestation is the carrying of young in the uterus.

Hydrothermal vents, found where the deep-sea floor pulls apart, spew up seawater and minerals superheated by contact with molten rock.

The **intertidal zone** is the area near or on shore, just above low-tide mark.

Invertebrates are animals that do not have backbones.

Liverwort is a class of plants that resemble moss.

Mollusks are invertebrate animals with a soft, unsegmented body enclosed in a shell.

The **Nazca** plate is a tectonic plate being pushed southeast where it is being driven beneath the South American continental plate.

Nocturnal animals are active mostly at night.

Order is a category for the classification of living things, below the class and above the family.

The **Pampa zone** refers to the highest elevation zone surrounding certain island summits. The climate and soil are moist and windswept, supporting water tolerant plants.

Pectoral corresponds to the location on the chest of an animal.

Photosynthesis is the chemical process by which plants turn light into food.

Plate tectonics is a theory which holds that a hard layer of the earth, composed of plates, rests on a softer more fluid layer below. As the plates move on this softer layer due to the Earth's heat energy, rifts, faults, and collisions between the plates create the features of the Earth, above and beneath the sea.

Predation is the act of an animal eating another animal.

The **Scalesia zone** is the area found on certain of the Galápagos Islands where the plant genus *Scalesia* thrives.

A **species** is a group of plants or animals that breed with one another and share physical characteristics.

Subspecies are populations of the same species genetically distinguishable by a particular geographical region.

Superheated temperatures are above the boiling point without converting into vapor.

Tuff is rock composed of fine volcanic detritus.

A **vegetarian** lives on a diet consisting of plants.

Vegetative zones are areas characterized by the predominance of certain vegetation.

Venomous indicates the ability to inflict a poisoned wound.

index

about the author

MARK BLUM has been photographing the natural world in 3-D for three decades, both as a stock photographer and on assignment. He has designed much of his own equipment to meet the special challenges of stereoscopic wildlife photography. This volume is Mark's fifth 3-D book for Chronicle Books, including *Beneath the Sea in 3-D*, *Bugs in 3-D*, and *Amphibians & Reptiles in 3-D*. His stereophotography has also been published in magazines, on CD-ROM, and on ViewMaster® reels under a Discovery Channel® license. In commemoration of the Year of the Ocean, Mark's underwater 3-D images were projected onto a specially installed 41-foot screen for thousands of viewers attending the Expo '98 World Fair in Lisbon, Portugal. Mark makes his home on the Monterey Bay in California. (Email: markb@redshift.com; Web site: www.Undersea3d.com)